THE ARTEMIS ROCKET

An evolution of lunar and space science

By

Russell L. Arthur

All rights reserved. No part of this publication may be reproduced, distributed, or transmitted in any form or by any means, including photocopying, recording, or other electronic or mechanical methods, without the prior written permission of the publisher, except in the case of brief quotations embodied in critical reviews and certain other noncommercial uses permitted by copyright law.

Copyright © Russell L. Arthur, 2022.

Table of contents

INTRODUCTION

Chapter 1: The Launching of the Rocket

Chapter 2: Its Missions

Chapter 3: How the rocket operates

Chapter 4: Things to know about NASA's moon-bound mega-rocket
- Artemis Facts

Chapter 5: Objects and surprises aboard the rocket

Chapter 6: The best places to obtain the finest view of the launch

Chapter 7: The future of rocket and lunar science
- WHAT'S NEXT?
- How SLS compares to Apollo-era Saturn V
- Artemis 1 vs. Apollo 4
- A sneak glimpse into Artemis 2

Frequently asked questions: Your curiosities acknowledged

INTRODUCTION

2022 marks half a century since Apollo 17 astronaut Eugene Cernan left the final footsteps on the moon in 1972, and a lot has happened since then.

That year, the first scientific hand-held calculator was launched; now we carry more computing capability in our pocket than that which successfully led the Apollo astronauts to the moon and back.

Now, at long last, mankind is poised to depart low Earth orbit (LEO) again. Only two dozen astronauts have accomplished that achievement so far, all of them white guys. Soon the first female astronaut and the first astronaut of color will join the renowned lists of moonwalkers.

It's all because of the Artemis mission, NASA's ambition to explore more of the lunar surface than ever before.

By 2025, we might witness astronauts walk in the lunar dust once again, and in far higher clarity owing to the enhancements from grainy black and white film footage that half a century of technological advancement has delivered. A whole new generation might regard themselves as budding space explorers, driven to dream big.

NASA's new moon rocket landed at the launch pad Wednesday ahead of its inaugural flight in less than a week.

The 322-foot (98-meter) rocket exited from its massive hangar late Tuesday night, bringing hundreds of Kennedy Space Center employees, many of whom were not yet alive when NASA took humans to the moon a half-century ago. It took almost 10 hours for the rocket to complete the four-mile voyage to the pad, drawing up around daybreak.

NASA is planning for an Aug. 29 liftoff for the lunar test trip. No one will be inside the crew capsule atop the rocket, only three mannequins teeming with sensors to detect radiation and vibration.

The capsule will sail around the moon in a distant orbit for a few weeks, before returning for a splashdown in the Pacific. The total flight should take six weeks.

The flight marks the first moonshot in NASA's Artemis program. The space agency is aiming for a lunar-orbiting voyage with humans in two years and a lunar landing by a human crew as early as 2025.

That's significantly later than NASA projected when it began the program more than a decade ago when the space shuttle fleet ended. The years of delays have added billions of dollars to the expense.

"Now for the first time since 1972, we're going to be launching a rocket that's built for deep space," NASA's rocket program manager, John Honeycutt, stated recently.

NASA's new SLS moon rocket, short for Space Launch System, is 41 feet (12 meters) shorter than the Saturn V rockets launched during Apollo a half-century ago. But it's more powerful, employing a core stage and twin strap-on boosters, akin to the ones used for the space shuttles.

"When you look at the rocket, it almost seems vintage. It appears like we're looking back toward the Saturn V," NASA Administrator Bill Nelson told reporters earlier this month.

"But it's a whole different, new, extremely advanced, more sophisticated rocket and spacecraft."

Twenty-four astronauts traveled to the moon during Apollo, with 12 of them landing on it from 1969 through 1972. The space agency wants a more diversified staff and more persistent effort under Artemis, named after Apollo's legendary twin sister.

"I want to underline that this is a test flight," Nelson added. "It's only the beginning."

This was the rocket's third trip to the pad. A countdown test in April was plagued by fuel leaks and other technical issues, requiring NASA to return the rocket to the hangar for repairs. The dress rehearsal was conducted at the pad in June, with better outcomes.

SLS is the first rocket intended to transport both humans and cargo on a single flight.

Chapter 1: The Launching of the Rocket

With a date and time for the return to moon launch fixed, the Space Coast is once again preparing for massive crowds to attend to observe history.

The first window for the Artemis 1 mission's SLS launch opens at 8:33 a.m. on Aug. 29.

"It's just the week before Labor Day weekend, so I expect at least 100,000. But probably much more than that," said Peter Cranis, the executive director of the Space Coast Office of Tourism.

While the cadence of launches from the Kennedy Space Center has ramped up considerably this year, the SLS will be the most powerful rocket to blast out there in years.

Already enjoying 15 months of record-breaking visitor numbers, hotels are receiving a late summer boost owing to the launch.

NASA officials said the last phase of pre-flight testing for the difficult Artemis I SLS rocket had some triumphs but a few obstacles remained.

The way the NASA and Artemis leadership laid it out in a post-test teleconference — they got through 90% of what they needed from June 20's wet dress rehearsal.

One of the great milestones they were able to achieve was fuelling up the different tanks of the Space Launch System rocket stack with hundreds of thousands of super-cooled fuel, an hours-long and delicate procedure.

But a hydrogen leak, and a few other difficulties, precluded a full run of the comprehensive and important pre-flight test.

The Artemis launch director declared it a terrific day as much was done.

"The bulk of our goals were attained. There were maybe little components inside that major target that we maybe came up a little bit short on," Charlie Blackwell-Thompson, NASA's Artemis Launch Director, said.

The fake launch countdown ticked down to T-29. Short of the aim to bring it close to T-9 seconds since the launch commit requirements were not satisfied. In other words, if the rehearsal was the actual launch day, it would have been scrubbed.

"Hopefully they learned a great lot and got it reasonably near to where they would have been able to have a successful countdown," said Don Platt of Florida Tech. "And they take the lessons learned and apply them, repair that 10% of issues."

The Space Launch System, whose construction throughout the last decade has been driven by Boeing Co (BA.N), emerged from its assembly building at NASA's Kennedy Space Center in Florida at 10 p.m. EDT (0200 GMT) on Tuesday and started a four-mile (6-kilometer) walk to its launchpad.

Moving less than 1mph (1.6kph), the rollout will take around 11 hours.

Sitting atop the rocket is NASA's Orion astronaut capsule, constructed by Lockheed Martin Corp (LMT.N). It is meant to split from the rocket in orbit, convey passengers toward the moon and meet with a separate spacecraft that will take men to the lunar surface.

For the Aug. 29 mission, named Artemis 1, the Orion capsule will launch atop the Space Launch System without any people and circle the moon before returning to Earth for an ocean splashdown 42 days later.

If adverse launch weather or a small technical problem forces a delay on Aug. 29, the National Aeronautics and Space Administration has backup launch dates on Sept. 2 and Sept. 5.

To honor the significant feat that's shortly to take place, NASA is going all out with celebrations pre-launch, day-of launch, and post-launch on-site and through a webcast.

The space agency has a list of star-studded guests for the event, including Jack Black, Chris Evans, and Keke Palmer, but it hasn't been confirmed whether they'll be attending in-person or digitally.

Additionally, Josh Groban and Herbie Hancock will sing "The Star-Spangled Banner" while the Philadelphia Orchestra and cellist Yo-Yo Ma will play a version of "America the Beautiful."

This spaceship, mostly developed by Lockheed Martin, will remain in orbit 'longer than any ship for humans has done without docking to a space station and come home quicker and hotter than ever before,' NASA has claimed earlier.

If Artemis I is a success, then in 2024 NASA will launch Artemis II on a voyage around the moon, this time with a human crew on board.

The Artemis II mission proposes to deploy four astronauts in the first crewed Orion spacecraft into a lunar flyby for a maximum of 21 days.

Chapter 2: Its Missions

The Artemis I mission is the first in NASA's series of "increasingly difficult" missions into space and around the moon. The flight test will be uncrewed and serve as a stepping stone for future human space travel.

Furthermore, the mission attempts to "show [NASA's] dedication and capabilities to extend human presence beyond the Moon and beyond," according to the space agency. The Space Launch System rocket and the Orion spacecraft are the two vehicles departing Earth.

The major goal of the SLS mission is to test the technology for a future voyage involving humans. But the rocket will also be carrying several payloads to assist research studies.

Among the payloads will be a CubeSat meant to seek water on the moon's surface. CubeSats are miniature research spacecraft also known as nanosatellites. They give a low-cost option for scientists, governments, and commercial groups to carry out space research.

NASA names the CubeSat onboard the SLS Lunar IceCube. It is roughly the size of a shoe box and weighs 14 kg.

While circling the moon, Lunar IceCube will use an instrument called a spectrometer to hunt for and analyze lunar ice. A spectrometer is a device used to measure atomic and molecular processes.

NASA has stated the hunt for ice on the moon is vital since water is a crucial resource for future exploration efforts. Astronauts may utilize the ice for drinking water and to chill equipment or create rocket propellants for voyages farther into the solar system.

The space agency stated in a recent statement that Lunar IceCube would be gathering data on the absorption and release of water from the regolith: the moon's stony and dusty surface. "With Lunar IceCube exploring this process, NASA can track these changes as they occur on the Moon," the release stated.

The CubeSat will also examine the ecosphere. The exosphere is the thin, outer region encircling a planet or satellite object, such as the moon.

NASA says data obtained by Lunar IceCube will offer scientists a better understanding of how water and other things behave on the moon. With this knowledge, researchers expect to be able to forecast seasonal variations influencing lunar ice that might influence its utilization as a resource.

"Humans provide a wonderful resource in traveling back to the surface of the moon," Debra Needham, a planetary scientist at NASA's Marshall Space Flight Center in Huntsville, Alabama, told.

"It's simpler for an astronaut to pick up a rock and then, out of the corner of their eye, notice another rock and quickly mark it as intriguing," Ryan Watkins, a planetary scientist at NASA's Jet Propulsion Laboratory, told. "Real-time decision making is a lot faster and more efficient."

The first landing site
To do the finest research possible, astronauts must travel to the proper spots on the moon. The six Apollo landing sites were concentrated in a zone on the near side of the moon. Artemis 3, which is supposed

to be the first Artemis mission to plant boots in the lunar regolith, will instead journey to the lunar south polar area.

Not only is this location intriguing for its huge volumes of water ice concealed in perpetually shadowed craters, but it is the first time astronauts will have gone there, far from the Apollo landing sites.

"While they were reasonably scattered on the surface, the Apollo missions still essentially traveled to one part of the moon that may have been impacted by the extremely huge impact that caused Mare Imbrium," Needham said.

Mare Imbrium ("Sea of Rains") is a massive, 712-mile-wide (1,146 kilometers) impact basin that was created 3.9 billion years ago and was later filled by lava.

Scientists have theorized that the construction of this impact basin corresponded with the Late Heavy Bombardment, which was postulated as a time between 4.2 billion and 3.9 billion years ago when the inner planets were being battered by asteroids and comets.

However, newer discoveries have placed doubt on this notion, and sampling other portions of the moon might assist to resolve the argument.

"One reason why we're moving so far away from that impact basin is to be able to reach various rock types that are perhaps older and retain a record of a more ancient moon," Needham said. "We'll be able to get a solid feel of whether there was a time of extremely intensive bombardment or not."

Lots of robotic science

Artemis isn't simply about astronauts. Although humans will be able to accomplish more research in a shorter period, plans for only one crewed flight per year suggest that astronauts won't be able to go everywhere on the moon.

So, to help lunar research during the Artemis mission, hundreds of robotic experiments will be sent to places throughout the moon between 2022 and 2025.

These studies are being enabled via the Private Lunar Payload Services (CLPS), in which commercial companies are cooperating with NASA on a $2.6 billion initiative to fly modest research missions to the moon, with 46 payloads completed so far.

Watkins said that her favorite payload is a probe called the Lunar Vulkan Imaging and Spectroscopy Explorer (Lunar-VISE), which will land on the Gruithuisen volcanic domes that lie between Mare Imbrium and Oceanus Procellarum ("Ocean of Storms") and that Watkins studied as a student using orbital data.

These domes are thought to be silicic, indicating they were created by silica-rich magma. But on Earth, silicic-based volcanic constructions need plate tectonics and occur around plate borders in the seas.

How, therefore, could they have developed on the moon, which has neither seas of water nor plate tectonics? There are several geological ideas, and Lunar-VISE will be equipped with both a stationary lander and a rover that will perform compositional measurements at different points around the domes to evaluate the various theories.

Water on the moon

Needham is especially enthused by several of the payloads in the CLPS program, including the Near Infrared Volatile Spectrometer System, and the Neutron Spectrometer System aboard a tiny robotic rover named MoonRanger, and the Mass Spectrometer Observing Lunar Operations.

Multiple versions of these three instruments will be transported to the moon by Pittsburgh-based Astrobotic Technologies later this year and by California-based Masten Space Systems in 2023.

They will function in tandem to detect and examine the range of "volatiles" — substances with low boiling points, such as water and carbon dioxide — on the lunar surface, in the near-subsurface, and just above the surface, in the thin "exosphere."

"What's exciting is that we're flying many copies of these payloads, with the initial iteration traveling to a nonpolar zone, and then we'll also send them to a polar region in a subsequent delivery," Needham said. "So we're going to be able to compare these two quite different sections of the moon using the same sensors."

At first impression, the moon doesn't appear moist, and analyses of the rock samples sent back to Earth by the Apollo missions revealed them to be bone-dry.

But aside from the ice hidden in the cold traps of permanently shadowed craters at the poles, water molecules (or at least hydroxyl molecules, which are a proxy for water formed from an oxygen atom joined to a single hydrogen atom, rather than two hydrogen atoms as in the case of water) have been observed migrating across the lunar surface near the day-night terminator.

These observations were initially made by NASA's Moon Mineralogy Mapper instrument on board India's Chandrayaan-1 lunar orbiter in 2009, and in 2020, the soon-to-be-retired Stratospheric Observatory for Infrared Astronomy (SOFIA), which flies in the back of a modified Boeing 747, confirmed the presence of water molecules near the surface of the moon, albeit in a very thin veneer.

"We now know there's water pretty much everywhere on the moon," Needham added. "But it's from a different source and in a different form than the water in the south polar zone.

These payloads can provide us some ground-truth to the distant observations, and that's critical for feeding into models explaining water circulation on the moon."

It's not yet apparent how viable the human spaceflight element of the Artemis lunar program is and if future crewed missions will land on the moon's surface beyond Artemis 3, considering that each launch will cost over $4 billion.

On the other hand, the comparatively smaller cost of the CLPS element of the Artemis mission will ensure the delivery of hundreds of scientific experiments throughout the moon, offering scientists an unparalleled amount of fresh data to analyze.

"We're tremendously looking forward to the discoveries that we're going to make in new areas on the surface of the moon," Needham added.

Orion has flown before, once, on a near-Earth test mission in 2014. But that utilized an existing commercial rocket to go into orbit. This impending trip is consequently the first thorough end-to-end assessment of the Artemis exploration gear.

Sensors within Orion will collect data on the deep-space radiation environment and other elements of the trip. And, not long after liftoff, 10 small CubeSats will launch from an adapter linking Orion to the SLS's upper stage.

These small spacecraft will undertake a range of duties, from seeking water ice on the moon to going to a near-Earth asteroid using a solar sail.

But the fundamental purpose of Artemis 1 is to show that SLS and Orion are ready to transport humans, which the pair will achieve in pretty short order if Artemis 1 goes well. NASA wants to fly Artemis 2, a crewed mission to lunar orbit, in 2024. Artemis 3 would then set people down near the moon's south pole in 2025 or 2026, in the first crewed lunar landing since the Apollo 17 mission in 1972.

A significant collaborator on the future expedition is Europe.

It is supplying the propulsion module that sits on the rear of Orion, propelling it into space.

"More than 10 nations in Europe have been working on this European Space Agency (ESA) contribution. It's a tremendously crucial time for us," remarked Siân Cleaver from aircraft company Airbus.

"The European Service Module is not just cargo, it's not just a piece of equipment - it's a very crucial part because Orion can't travel to the Moon without us."

Europe believes its contribution to this and future SLS/Orion flights will finally see a European citizen get to be part of a lunar surface crew at some time.

For now, it will have to cheer for the British cartoon figure Shaun the Sheep. A puppet used in the stop-motion TV films has been installed in the Orion spacecraft, replete with an Esa insignia and Union flag on its overalls.

While Nasa is creating the SLS, the American rocket entrepreneur Elon Musk is preparing an even bigger spacecraft at his R&D center in Texas.

He names his big rocket the Starship, and it will play a part in future Artemis missions by hooking up with Orion to transport men down to the surface of the Moon.

Like SLS, Starship has yet to make the first flight. Unlike SLS, Starship has been planned to be entirely reusable and ought thus to be substantially cheaper to operate.

A recent evaluation from the Office of Inspector General, which monitors Nasa activities, revealed that the first four SLS missions would each cost more than $4bn to execute - an amount of money that was defined as "unsustainable".

The agency claimed adjustments made to the way it contracts industries will cut down future production costs considerably.

SLS - the world's highest rocket – is significantly more advanced than the Saturn V, technologically. But its principal aim is to push.

SLS will generate 8.8 million pounds (3.9 million kg) of thrust during liftoff and ascent, 15% more than the Saturn V. It'll take that much force to propel a spacecraft weighing roughly 6 million pounds (2.7 million kg) to orbit.

Propelled by a pair of five-segment boosters and four RS-25 engines, the rocket will reach the period of highest atmospheric force within 90 seconds, NASA claims. After jettisoning its rockets, service module panels, and launch abort system, the core stage engines will shut down.

At that moment, the core stage will separate from the Orion spacecraft.

The Orion moonship is known formally as the Orion Multi-Purpose Crew Vehicle or Orion MPCV. It'll proceed to Earth orbit atop of SLS after launch. There, it'll deploy its solar arrays and the Interim Cryogenic Propulsion Stage (ICPS) that'll give the ship the massive push required to leave Earth's orbit and go toward the moon.

Orion had performed an Earth-orbiting test in 2014, so this isn't its first excursion to space. But it is its first mission to the moon, and it'll get there by propulsion by a service module supplied by the European Space Agency.

The service module will provide the spacecraft's primary propulsion engine and electricity (as well as house air and water for astronauts on future missions).

Orion will go as near to the moon's surface as around 62 miles (100 km). It'll utilize the moon's gravity to drive itself into an orbit around 40,000 miles (70,000 km) from the moon.

The spacecraft will linger in that orbit for around six days, gathering data. During that time, mission controllers will analyze its performance.

Then it'll execute a second close flyby of the moon, approaching within roughly 60 miles (100 km). Another perfectly timed engine firing of the European-provided service module — in concert with the moon's gravity — will propel the moonship back toward Earth.

It'll reach our planet's atmosphere moving at 25,000 mph (11 km/second), causing temperatures of roughly 5,000 degrees Fahrenheit (2,760 degrees Celsius). So it'll fly faster – and be hotter – than during its 2014 flight test.

In total, the mission will take around a month and traverse a distance of 1.3 million kilometers. It's anticipated to make a flawless landing within view of the rescue ship off the coast of Baja, California.

Chapter 3: How the rocket operates

Two enormous solid rocket boosters and a core stage laden with 733,000 gallons (2.8 million liters) of fuel will push Orion through Earth's atmosphere to orbit. The SLS upper stage, called the Interim Cryogenic Propulsion Stage (ICPS), will then burn to launch Orion toward the moon.

Once the ICPS has fired, it has another job: to launch 10 miniature CubeSats that are hitch-hiking onboard Artemis 1. These small spacecraft include BioSentinel, a mission that will take yeast samples beyond LEO.

The objective is to examine radiation levels and their impact on living beings, which will give crucial insights into keeping astronauts safe when they travel on future Artemis missions.

After separation from the ICPS, Orion will be pushed and powered by the European Service Module, developed by the European Space Agency (ESA). "The Service Module will also offer consumables for the future crew, including water and oxygen," stated Phillippe Berthe, ESA's project coordination manager for the module.

When liquid hydrogen and oxygen are injected into the engine chambers and ignited with a spark, the chemical reaction creates huge quantities of energy and steam.

The steam leaves engine nozzles at rates of 16,000 km/h (10,000 mph) to create thrust - the power that propels a rocket into the air.

The SRBs provide the rocket with more strength to escape gravity's grasp. These twin rockets tower more than 17 stories tall and burn six tonnes of solid propellant per second. They deliver 75% of the total thrust during the first two minutes of flight.

Artemis 1 will be gone for between 26 and 42 days. It'll take one to two weeks to reach the moon, where Orion will swoop down near to the lunar surface and utilize the gravitational boost it gets to enter a so-called "distant retrograde orbit."

Retrograde indicates that it will circle the moon in the opposite direction to the one in which the moon rotates. Orion will linger in that orbit for between six and 19 days. Then it will swing back down toward the moon for another kick to assist propel its nine- to 19-day trek back to Earth.

So how does the new Orion service module compare to the lunar modules that took Apollo astronauts to the moon? "The propulsion is substantially the same; it is pretty analogous to the Apollo period," stated Berthe.

Yet half a century of technological advancement has brought additional steps ahead. "There have been enormous breakthroughs in solar cells," Berthe added.

"Computing power is another important advance," added Berthe. The Apollo astronauts famously went to the moon with less computational power than available in an iPhone. That means a lot of manual chores for the staff. This time around, the spacecraft's sophisticated computers can perform most of the hard work.

The most powerful rocket ever?

If we use thrust as a metric, the SLS will be the most powerful rocket ever when it travels to space in 2022. The Block 1 SLS will produce 8.8 million pounds (39.1 Meganewtons) of thrust during launch, 15% more than the Saturn V.

In the 1960s, the Soviet Union created a rocket dubbed the N1, in an attempt to reach the Moon. Its first stage could provide 10.2 million pounds (45.4 Meganewtons) of thrust. But all four test flights ended in disaster.

A future version of the SLS - termed Block 2 cargo - should reach the N1's thrust levels. But a spacecraft named Starship, being constructed by Elon Musk's firm SpaceX, should surpass both - generating as much as 15 million pounds (66.7 Meganewtons) of thrust. Starship is presently in development, however, there is no fixed schedule for its maiden orbital flight.

The SLS in numbers

The rocket will stand 98m (322ft) tall in its original, or Block 1, Configuration

The Block 1 SLS can transport more than 27 tonnes (59,500 pounds) to lunar orbits - the equivalent of 11 big sports utility vehicles (SUVs)

A future version of the SLS, designated Block 2 Cargo, will launch 46 tonnes (101,400 pounds) to the Moon. That's 18 huge SUVs.

The SLS will generate 8.8 million pounds (39.1 Meganewtons) of thrust in its Block 1 configuration

Four RS-25 engines lie at the foot of the core stage; they're the same ones used in the space shuttle

How shuttle technology was re-used

The SLS core stage is based on the space shuttle's foam-covered exterior tank. This tank provided propellant to three RS-25 engines in the back of the shuttle orbiter. The solid rocket boosters have substantially the same job in both vehicles.

But the SLS is a different beast. Several components and structures originating from the shuttle received major design adjustments because of the varying amounts of stress exerted on them by the SLS.

As an illustration of these various strains, in the space shuttle, the RS-25 engines were canted up and away from the solid rocket boosters. Moving them near to the SRBs exposes them to additional shaking.

As a consequence, every system in the complicated SLS engine section had to be extensively tested to guarantee it could resist vibrations.

Why the SLS was created

In February 2010, the Obama administration scrapped Constellation - George W Bush's controversial plan to return to the Moon by 2020.

The revelation came as a severe blow to employees in five southern states - Alabama, Florida, Louisiana, Mississippi, and Texas - where Nasa's human spaceflight project subsidized tens of thousands of employees.

Some Capitol Hill lawmakers were outraged. At the time, Richard Shelby, a republican senator from Alabama, stated Congress would not "sit back and witness the reckless abandoning of fundamental principles, a proven track record, a steady route to success, and the ruin of our human spaceflight program".

As a compromise, legislators from impacted states insisted on a single super heavy-lift rocket to replace the Constellation launchers canceled by the White House.

The SLS design was debuted in 2011. After development began, delays and cost overruns supplied ammo to detractors, who argued Nasa should depend on rockets controlled by private suppliers.

But without considerable changes, no current launchers have the capacity to deliver Orion, humans, and big cargo to the Moon in one mission - as the SLS would have.

An estimated $18bn has been spent on the SLS from the beginning of the previous decade.

But with the rocket's construction phase done and its flight certification testing complete, the first SLS is currently at Florida's Kennedy Space Center, getting prepped for its inaugural launch.

John Shannon, who has been in charge of the SLS at Boeing since 2015, explained: "I anticipate that once SLS is in the national capabilities there won't be a need for another heavy-lift vehicle like it for many years. So this is truly a once-in-a-generation chance."

The Space Launch System's essential characteristics

Solid Rocket Boosters (SRBs)
Once the instruction to ignite the SRBs is provided by the onboard computer, a

booster charge shoots down the length of the rocket, which in turn ignites the solid rocket fuel.

Core Stage

Contains the liquid fuel that feeds the four RS-25 engines. It burns almost 2 million liters of liquid hydrogen (orange tank) and nearly 750,000 liters of liquid oxygen (blue tank) in eight minutes to reach low-Earth orbit.

Four RS-25 engines

Originally developed for use on the Space Shuttle, they provide power for the whole of the rocket's ascent and include a 'gimbal bearing', which maintains the rocket on track.

Orion spacecraft's major characteristics

Orion stage adaptor

A ring-shaped structure that holds the Orion spacecraft linked to the top of the assembly.

Interim Cryogenic Propulsion Stage (ICPS)

The ICPS is used to push the Orion capsule towards the Moon after it reaches low-Earth orbit. It employs liquid hydrogen and oxygen.

How large is NASA's Space Launch System?

The SLS Block 1 measures 98m high, while the SLS Block 2 will reach up to 111m high.

Chapter 4: Things to know about NASA's moon-bound mega-rocket

NASA recently moved its enormous heavy-lift rocket to a launchpad at Kennedy Space Center for some essential testing ahead of its first lunar trip.

It's been a long time since the U.S. space agency had a rocket of this caliber, capable of transporting substantial payloads - humans and cargo — into deep space.

Not only is the Space Launch System, or SLS, constructed to fly to the moon, it's projected to one day put millions of miles on the odometer during the first crewed mission to Mars. Robotic research expeditions to Saturn and Jupiter possibly potentially be in its future.

Here are some crucial facts about the mega-rocket as it prepares for its inaugural journey, the Artemis I mission to lunar orbit.

1. It's the only rocket that can launch the Orion spacecraft to the moon

SLS is the only rocket capable of delivering the Orion spacecraft, a capsule that sits atop the stack of rockets, to the moon and beyond.

Think of the Orion capsule as the RV of the sky: It's not simply a transport but a residence for up to four people. To undertake extended excursions into deep space, individuals will need to be able to eat, sleep, work, and spend time onboard for months.

For Artemis I, an uncrewed Orion will travel thousands of kilometers over and around the moon. Three weeks after liftoff, the capsule will splash down in the Pacific Ocean.

The goal of the maiden Artemis mission is to test its capacity to safely reenter Earth's atmosphere and drop into the precise area for the Navy to retrieve.

2. It's not the size, but the push, that matters

Standing 322 feet high, the mega-rocket is higher than the Statue of Liberty and London's Big Ben. Compare that to the 184-foot Space Shuttle rocket, which launched humans to the space station in low-Earth orbit.

Despite towering above its predecessor, SLS is a touch shorter than Saturn V, the last rocket NASA used to transport humans into deep space. The Apollo-era rocket was 41 feet higher.

But the new rocket is undeniably more powerful. SLS will create 8.8 million pounds of thrust — the oomph an engine produces for the rocket — during liftoff and ascent. That's 15 percent more than Saturn V provided. Future variations of the new rocket will carry even greater power.

The four primary SLS engines, fuelled with 700,000 gallons of cryogenic, or ultra-cold, propellant, will deliver a thrust strong enough to keep eight Boeing 747s aloft.

3. The mega-rocket is state-of-the-art 1980s technology

SLS is physically and symbolically built upon the Space Shuttle heritage. NASA included important components of the shuttle, which operated between 1981 and 2011, into the new rocket.

Engineers changed the famous space aircraft out for either a cargo or Orion crew ship. The middle orange core is an extended shuttle external fuel tank, driven by four shuttle engines.

Rather than recycling those engines, however, NASA will drop them in the ocean. Twin shuttle solid rocket boosters will support the core during the first portion of the mission, giving 75 percent of the first upward push.

It's not all ancient tech, however. NASA improved certain hardware and employed new tools and production procedures to get the task done. Some sections have been modified to match the demands of deep-space flight, but Congress didn't authorize the space agency to start totally from scratch to construct the current mega-rocket.

4. Sorry, environment. It's not reusable.

Remember that the new moon rocket is made utilizing shuttle components. NASA developed the shuttle to ferry personnel and supplies back and forth to the space station, which orbits about 250 miles from Earth.

To alter the rocket so that it could fly far further into space, researchers needed to decrease the burden. After all, the moon is around 239,000 miles from Earth, nearly

1,000 times the distance of the space station.

Engineers removed the Shuttle's reusable rockets, parachutes, backup fuel, and landing sensors from the design - the system that enabled the agency to deploy it again.

This gives NASA back around 2,000 pounds of additional weight capacity for lunar flights. Doing so will help Orion achieve 24,500 mph, the speed required to launch it on a moon-bound course.

But this implies SLS will require fresh rockets for each flight.

At least the engine exhaust is somewhat "clean," superheated water vapor. The engines are given liquid hydrogen and liquid oxygen fuel. And NASA updated the rocket insulation from asbestos to rubber materials, also an environmental advantage.

5. The mega-rocket has an all-American price tag

Many officials at NASA and in Congress refer to SLS as "the nation's rocket," the "flagship rocket," or "America's rocket."

It's considered a national asset, somewhat unlike a tailored aircraft carrier for the military, meant to serve a national interest: exploring the solar system.

That's the key reason it's regarded to be the most costly rocket ever constructed. While the expanding commercial spaceflight industry may soon show it can develop a more cost-efficient space transportation system, the price was never the aim for SLS.

When Congress issued a NASA appropriations bill in 2010, it authorized the space agency to construct the rocket, even specifying what components to use, which businesses to hire, and what type of commercial ties to exploit. At that time, during the Great Recession, the politicians tried to sustain thousands of jobs in their districts. Artemis is not simply a space initiative, but a jobs program.

About 3,800 vendors in all 50 states have contributed to the rocket and Orion missions, said Tom Whitmeyer, NASA's deputy associate administrator for shared exploration systems.

"When you see this rocket, it's not simply a piece of metal that's going to sit at the pad. It's a big lot of individuals, rocket scientists around our nation, throughout our agencies, who have worked on this."

"It's a symbol of our nation and our towns, our aerospace sector, and what's in collaboration behind it," he said in a teleconference with reporters in March.

"When you see this rocket, it's not simply a piece of metal that's going to sit at the pad. It's a big lot of individuals, rocket scientists around our nation, throughout our agency, who have worked on this."

At a March congressional committee, Inspector General Paul Martin, who acts as the space agency watchdog for the federal government, projected each launch would cost $4.1 billion, with half of the expense ascribed simply to SLS. For context, that's approximately one-fifth of the whole NASA budget.

By 2025, Martin believes NASA will have spent $93 billion on the Artemis program.

6. The rocket is the ultimate Transformer

Engineers planned SLS to grow into greater powerful versions as its Artemis missions get more sophisticated.

The initial assembly, designated "Block 1," will employ the middle (orange) core booster with four main engines. It can launch over 59,500 pounds to orbit beyond the moon.

Additionally, a pair of solid rocket boosters and liquid fuel-fed engines will generate most of its thrust. After exiting Earth's atmosphere, the last rocket booster — the Interim Cryogenic Propulsion Stage — pushes the Orion spacecraft forward to the moon.

This is the configuration NASA expects to employ for the first three Artemis missions, including a lunar landing.

Later flights, which will transport people, will have a different rocket architecture, including the strong Exploration Upper Stage. Known as "Block 1B," this rocket type can carry people and massive quantities of cargo - up to 83,700 pounds.

The next incarnation of SLS, called "Block 2," will deliver 9.5 million pounds of thrust and will be the workhorse vehicle for carrying cargo to the moon, Mars, and other deep-space destinations, an eight percent increase over Artemis I. This rocket will lift a massive 101,400 pounds.

In the inhospitable areas NASA astronauts are heading, they'll require loads of supplies.

Artemis Facts

SLS generates 8.8 million pounds of maximum thrust, 15 percent more than Apollo's Saturn V rocket.

Artemis I's SLS can deliver a payload into deep orbit weighing up to 83,700 pounds.

The SLS's 212-foot-tall core stage is produced by the Boeing Company at a NASA site in New Orleans.

The core stage can contain 730,000 gallons of fuel for its four RS-25 engines.

It was test fired at NASA's Stennis Space Center in Mississippi.

SLS features two solid rocket boosters positioned on either side of the core. Built by Northrop Grumman, they're adapted from solid rocket boosters that were built for NASA's Space Shuttle program.

SLS is planned to adapt for numerous destinations beyond the moon, including Mars, and robotic missions to Saturn and Jupiter.

An Artemis Base Camp on the surface and a gateway in lunar orbit will be built up by NASA. This will assist in enabling exploration by robots and astronauts. The gateway is a crucial component since it will function as a multipurpose outpost circling the moon.

Other space agencies will also contribute to the Artemis mission.

Canadian Space Agency will contribute by supplying sophisticated robots for the gateway.

The European Space Agency will supply the International Habitat and the ESPRIT module, which will give extra communications capabilities, among other things.

Japan Aerospace Exploration Agency proposes to provide habitation components and logistical replenishment.

Chapter 5: Objects and surprises aboard the rocket

While no human crew will go onboard NASA's Artemis I mission, it doesn't imply the Orion spacecraft will be vacant.

When the Space Launch System rocket and Orion capsule, slated for liftoff on August 29, set off on a mission beyond the moon, the spacecraft will be carrying some unique objects on board.

Inside Orion will be three mannequins, toys, and possibly an Amazon Alexa, along with historic and educational artifacts.

The mission — which will begin off the Artemis program, to someday return people to the moon — follows a tradition that began in the 1960s of NASA spacecraft bringing souvenirs. The tradition includes the Voyager probe's gold record and the Perseverance rover's microchip of 10.9 million names. Artemis I will carry 120 pounds of keepsakes and other stuff in its approved flying kit.

Moonikins reporting for duty

Sitting in the commander's seat of Orion will be Commander Moonikin Campos, a fitted mannequin that can gather data on what future human crews would experience on a lunar voyage.

Its name, decided through a public contest, is a homage to Arturo Campos, a NASA electrical power systems manager who helped in Apollo 13's safe return to Earth.

The commander's station has sensors in place under the seat and headrest to detect acceleration and vibration for the length of the mission, which is planned to last roughly 42 days. The mannequin will also wear the new Orion Crew Survival System suit meant for astronauts to wear during launch and descent. The suit contains two radiation sensors.

Two "phantoms" called Helga and Zohar will travel in other Orion seats. These mannequin torsos are comprised of materials that replicate the soft tissue, organs, and bones of a woman.

The two torsos feature more than 5,600 sensors and 34 radiation detectors to assess how much radiation exposure happens throughout the operation.

The mannequins are part of the Matroshka AstroRad Radiation Experiment, a partnership between the German Aerospace Center, the Israel Space Agency, NASA, and institutions across many nations.

Zohar will wear AstroRad, a radiation protective gear, to evaluate how effective it may be if future astronauts experience a solar storm.

Amazon's Alexa will be aboard for the voyage as a technological showcase built with Lockheed Martin, Amazon, and Cisco.

The tech experiment, named Callisto, comprises customized versions of Alexa, Amazon's voice assistant, and Cisco's teleconferencing platform Webex to evaluate how these apps work in space.

The purpose of Callisto, named after one of Artemis' hunting companions from Greek mythology, is to illustrate how astronauts and flight controllers may employ technology to make their tasks safer and more efficient as humans explore deep space.

Callisto will ride along on Orion's center console. The touch-screen tablet will communicate video and audio live between the spaceship and Johnson Space Center's Mission Control Center in Houston.

Toys in Space

Snoopy and space simply go together. The iconic figure created by Charles M. Schulz has been connected with NASA missions since the Apollo program when Schulz wrote comic strips picturing Snoopy on the moon.

The Apollo 10 lunar module received the moniker "Snoopy" because its duty was to snoop about and survey the Apollo 11 landing location on the moon, according to NASA.

Snoopy will function as Artemis I's zero-gravity indicator.

A Snoopy plush first traveled to space in 1990 on the Columbia spacecraft.

A pen nib used by Schulz from the Charles M. Schulz Museum and Research Center in Santa Rosa, California, will accompany the Artemis I mission, wrapped in a space-themed comic strip. And a plush Snoopy doll will fly as a zero gravity indication in the capsule.

The CIA has a long history of employing toys in orbit as zero gravity indicators — so-called because they begin to float after the ship has achieved zero gravity.

As part of NASA's partnership with the European Space Agency, which donated the service module for Orion, a little Shaun the Sheep toy will also be an Artemis passenger. The character is part of a children's television spinoff from the "Wallace and Gromit" series.

Shaun the Sheep is depicted in front of a replica of the Orion spacecraft.

Four Lego Minifigures will also travel on Orion as part of a continuing partnership between NASA and the Lego Group, to engage schoolchildren and adults in STEM (science, technology, engineering, and math) education.

A space-time capsule

The Artemis I Official Trip Kit, which comprises hundreds of objects, carries a range of patches, pins, and flags to share with individuals who contributed to the initial flight after the capsule splashes down in the Pacific Ocean in October.

A number of the artifacts — such as space science badges from the Girl Scouts of America, digitized student dreams of lunar exploration from the German Space Agency, and digital submissions from the Artemis Moon Pod essay contest — commemorate the achievements of students and instructors with an interest in STEM.

A variety of trees and plant seeds will be on board as an homage to a similar practice that started on the Apollo 14 mission.

The seeds were eventually planted and become "Moon Trees" as part of an experiment to investigate the impact of the space environment on seeds. NASA will share the Artemis seeds with teachers and educational groups after the spacecraft returns.

Several Apollo relics are traveling for the voyage, including an Apollo 8 commemorative medallion, an Apollo 11 mission patch, a bolt from one of Apollo 11's F-1 engines, and a tiny moon pebble gathered during Apollo 11 that also traveled on the last space shuttle flight. The pieces were provided by the National Air and Space Museum, which will include them in an exhibit whenever they return.

Cultural artifacts will be aboard the aircraft, too.

A 3D-printed figure of the Greek goddess Artemis will accompany the space mission and eventually be on exhibit at Greece's Acropolis Museum. The European Space Agency issued a postcard of Georges Méliès' renowned "A Trip to the Moon" artwork for the flight kit.

And the Israel Space Agency contributed a stone from the lowest dry land area on Earth, the edge of the Dead Sea, to fly on Artemis 1, a voyage that will journey deeper than any person has gone before.

Chapter 6: The best places to obtain the finest view of the launch

You're certain to have a great view of the rocket launch anywhere on the Space Coast – so long as the weather and cloud cover cooperates, of course.

But if you're not from the Brevard County region and you're just visiting, these are the cities you should include in your search query and the approximate distance away from Kennedy Space Center.

"On the mainland"

Brevard is a lengthy county with 72 miles of shore. Watching a rocket launch at the beach is popular for visitors and residents. If you're "on the mainland" here, it implies you're blocks or miles from the shore. However, the Banana River and Indian River Lagoon flow through Brevard, and practically all of these communities in this section below provide a riverbank vista appropriate for a launch.

- Mims, 18 kilometers distant
- Titusville, 14 miles distant
- Port St. John, 12 miles distant
- Cocoa, 12 kilometers distant
- Merritt Island, 16 miles distant
- Viera, 27 kilometers distant
- Melbourne, 35 to 45 kilometers distant
- West Melbourne, 43 miles distant
- Palm Bay, 39 to 48 kilometers distant

If you're staying "on the beach ..."

Watching a rocket launch from the beach is very Space Coast. These are the coastal cities you should use in your search searches and the estimated miles from Kennedy Space Center.

- Cape Canaveral, 15 miles distant
- Cocoa Beach, 20 miles distant
- Satellite Beach, 30 miles distant
- Indian Harbour Beach, 32 miles distant
- Indialantic, 36 kilometers distant
- Melbourne Beach, 37 miles distant

Cities beyond the Space Coast

If you don't mind a trip, here's a list of cities (and counties) outside of the Space Coast to include in your search searches for a place to stay. Here's a tip: Leave early to enable yourself plenty of time to locate parking and get comfortable.

Volusia County, north of Brevard

Oak Hill, 35 miles distant, around 40 minutes

New Smyrna Beach, 49 miles distant, around an hour

Daytona Beach, 63 miles distant, around an hour and 10 minutes

Indian River County, south of Brevard

Sebastian, 66 to 71 miles distant, around an hour and 15 minutes

Vero Beach, 69 to 81 miles distant, roughly 75 to 90 minutes

Wabasso, 61 to 73 miles distant, around 70 to 90 minutes

St. Lucie County, south of Brevard

Fort Pierce, 87 to 94 miles distant

Jensen Beach, 102 to 108 miles distant, 90 minutes to two hours

Martin County, south of Brevard

Stuart, 116 miles distant, roughly two hours

Hobe Sound, 131 miles distant, roughly two hours

Orange County, west of Brevard

East Orlando, 45 to 60 miles distant, around an hour

University of Central Florida, 35 miles distant, around 50 minutes

When should I arrange to see Artemis launch?

NASA has three tries for its SLS rocket launch:

Launch window 1: 8:33 to 10:33 a.m. EDT Monday, Aug. 29

Launch window 2: 12:48 to 2:48 p.m. EDT Friday, Sept. 2 (Labor Day weekend)

Launch window 3: 5:12 to 6:32 p.m. EDT Monday, Sept. 5 (Labor Day)

While you're on the Space Coast

While you're here to witness NASA's SLS giant moon rocket launch, may as well make it a mini-vacation.

There are lots of things to do and places to go. For example, the Space Coast offers 72 miles of coastline, and some of the finest surfers in the world grew up in this region. Coincidentally, the 37th annual National Kidney Foundation Rich Salick Surf Festival will be on Sept. 3-5 at Cocoa Beach.

Aside from Kennedy Space Center, the Space Coast is home to Port Canaveral, the second busiest port in the world, and Brevard Zoo.

Places to stay include holiday rentals, beds, and breakfasts, condo rentals, campsites, and hotels.

Florida vacation rentals

Sites like Airbnb, Vacation Rental By Owner, hotels.com, search engines like Google and Bing, or third-party travel-booking sites contain availability details.

Campgrounds in Brevard County and Central Florida

Making the trek to the Space Coast with a camper? Consider these possibilities below from Brevard County Parks & Recreation:

Manatee Hammock, a 26-acre campsite in Titusville 8 miles distant from Kennedy Space Center

Wickham Park, a 391-acre campsite near Melbourne 31 miles distant from Kennedy Space Center

Long Point Park, an 84-acre campsite at Melbourne Beach 52 miles distant from Kennedy Space Center

There are, of course, several campsites just outside the Space Coast.

Sebastian Inlet State Park near Melbourne Beach is 51 miles distant from Kennedy Space Center.

Donald McDonald Park in Sebastian is 71 miles distant from Kennedy Space Center.

Chapter 7: The future of rocket and lunar science

WHAT'S NEXT?

After Artemis 1, assuming everything goes according to plan, a second trip — Artemis 2 — will launch and bring humans around the moon and back in 2024. Then in 2025 or 2026, Artemis 3 will see humans land on the moon near the lunar south pole.

Eventually, the space between the Earth and the moon might be crowded with spaceships ferrying products and humans back and forth. Jeff Bezos, the founder of Amazon and the rocket business Blue Origin, has indicated that the moon may be a site to put our heavy industries.

Doing so would free up living space on Earth and shift our environment-polluting equipment to where there isn't even an atmosphere, the notion goes.

The moon is also a suitable staging station for further solar system research, astronomers believe. The magnitude and scale of the SLS indicate exactly how hard we have to struggle to escape from Earth's gravitational constraints.

The moon's gravity, which is six times less than ours, is substantially simpler to run from. There are also large volumes of water on the moon. As water is H2O, that indicates an ample supply of oxygen.

The moon's top layer alone holds enough oxygen to support 8 billion humans for 100,000 years, experts have found. Liquid oxygen is also rocket fuel, therefore moon mining may lead to the establishment of off-Earth "gas stations" where voyaging spacecraft could fill up their tanks.

That's why Artemis Base Camp will be on the moon's south pole: We already know that there's plenty of water there. Lunar Flashlight, one of the tiny spacecraft hitching a ride on Artemis 1, will circle the moon and shoot infrared lasers into permanently darkened craters near the lunar poles to further disclose the amount and accessibility of water ice there.

The sunlight near the lunar south pole is also beneficial; it is lighted roughly 90% of the time, compared to two weeks of daylight followed by two weeks of darkness on the rest of the moon. That's fantastic news for a lunar base powered by solar panels.

The combination of these two elements — water and sunshine — may lead to a period when rocket ships frequently fill up close to Artemis Base Camp and blast off for more distant locales like Mars and the asteroid belt.

Former NASA administrator Jim Bridenstine undoubtedly views moon exploration as a critical step on our road toward becoming an interplanetary species.

He has argued that mankind needs "many years in orbit and on the surface of the moon to gain operational confidence for undertaking long-term work and maintaining life away from Earth before we can start on the first multi-year human journey to Mars."

It's all part of returning to where we came from. The iron in your blood and the calcium in your bones were formed within stars that hurled these elements throughout the cosmos when they perished.

Eventually, those atoms found themselves inside intelligent beings that dreamt of going between the stars and constructed cathedral-sized rocket ships to transport them there.

The Artemis 1 launch this year may just be a tiny step, but it's a crucial one. Future historians might look back on it as the moment mankind made a huge leap in its return to the moon, this time for good.

How SLS compares to Apollo-era Saturn V

NASA is prepared to fly the Space Launch System on the Artemis 1 test mission on Aug. 29. The Saturn V launched for the first time in 1967 at the adjacent Kennedy Space Center launchpad 39A.

The last time the space agency completed a test flight of a moon rocket was the massive Saturn V in 1967. Two years later, a Saturn V rocket would launch the Apollo 11 mission, delivering humans to the moon.

Now, NASA is preparing to launch the SLS and Orion spacecraft from Kennedy Space Center in Florida under the Artemis program, named after Apollo's twin sister in Greek mythology.

NASA's Space Launch System (SLS) rocket with the Orion spacecraft onboard is visible at dawn atop the mobile launcher as it arrives at Launch Pad 39B, Wednesday, Aug. 17, 2022, at NASA's Kennedy Space Center in Florida.

NASA's Artemis I flight test is the first integrated test of the agency's deep space exploration systems: the Orion spacecraft, SLS rocket, and supporting base systems. The launch of the uncrewed flight test is slated for no early than Aug. 29.

NASA is aiming Aug. 29 for liftoff sending the Orion spacecraft on the Artemis-1 test trip, a 42-journey around the moon and back. No astronauts will be on board the test trip. However, if all goes well, four humans will embark on Artemis-2, circling the moon before the Artemis-3 lunar landing in 2025.

The SLS and Saturn V have major differences and a few shared characteristics.

Both SLS and Saturn V may launch from KSC launchpad 39B. The Saturn V utilized pads A and B at Launch Complex 39.

Pad A is presently leased by SpaceX, where the business launches NASA astronauts to the International Space Station and cargo for clients. SLS's designated launchpad will be Pad B.

SLS and Saturn V were both stacked inside the Kennedy Space Center Vehicle Assembly Building (VAB) before being moved on movable launch towers to the launch facility.

The fundamental difference between the moon rockets is power and how the vehicles are pushed off the earth. Side-by-side, Saturn V is taller, towering at 363 feet, while the Artemis-1 version of the SLS rocket is 322 feet tall.

About 75% of the SLS's thrust originates from two solid rocket boosters (SRBs) on each side of the core stage. When the SLS launches, it will utilize the SRBs and four RS-25 rocket engines to provide 8.8 million pounds of thrust to get off the Earth.

The three-stage Saturn V featured 5 F-1 engines on its first stage, providing 7.5 million pounds of force to take off. Once in orbit, five engines on the second stage supplied the power to propel the Apollo on its mission to the moon.

One of the things the two rockets have in common is the propulsion systems are both created by Aerojet Rocketdyne and the company's predecessors. The business supplies 39 propulsion parts for the SLS and Orion from the booster, upper stage, the jettison motor for the launch abort system, and Orion's primary engine.

The RS-25 engine is the main booster engine for SLS and was previously employed during the Space Shuttle Program. The rocket engine was conceived and constructed in Los Angeles before being assembled and tested at NASA's Stennis Space Center in Mississippi.

The SLS's upper stage engines, the RL-10 fuelled by liquid hydrogen and liquid oxygen, are built in West Palm Beach, Florida.

"The RL-10 will drive the top stack of the Artemis vehicle over 20,000 mph to depart Earth's gravity and make its journey to the moon," Maus stated.

Artemis 1 vs. Apollo 4

The Earth was shot by Apollo 4 from a distance of 11,214 miles.

The initial test flights for NASA's big moon rockets have some comparable test goals. However, the missions for Orion and Apollo 4 will appear extremely different.

Goals for the Artemis 1 test mission include evaluating Orion's communication, navigation, and guidance systems, assuring the overall launch and performance of SLS, and confirming Orion's heat shield can endure Earth re-entry at 25,000 mph. Many of them were the same for Apollo 4, the first Saturn V launch.

The Artemis 1 test flight will take Orion on a 42-day voyage around the moon. In 1967, Saturn V launched Apollo 4 on an 8-hour high elliptical orbit of Earth.

At its maximum distance, Apollo 4 was 11,234 miles from Earth. This was done so that when the spacecraft returned, it would imitate a lunar re-entry. The Orion will fly nearly 40,000 miles on the opposite side of the moon, further than any human-rated spaceship.

Apollo 4 concluded the trip by splashing down in the Pacific Ocean, and Orion will do the same. Both spacecraft suffers re-entry velocities over 25,000 mph, which is why the heat shield is so crucial.

A sneak glimpse into Artemis 2

Artemis 2 is the second planned flight of the Artemis program.

Following the launch and splashdown of the uncrewed Artemis 1, Artemis 2 intends to send people into the moon's orbit for the first time since 1972.

Artemis 2 will employ the massive Space Launch System (SLS) mega rocket and Orion spacecraft to launch the astronauts on an eight-day journey. The astronauts and mission controllers will gather data about Orion and the crew's performance to judge how ready the Artemis program is to take humans to the moon's surface.

The crew for Artemis 2 has not yet been named. The launch date is provisionally slated for 2024, given that all data from Artemis 1 shows that Artemis 2 is ready for flight. Assuming Artemis 2 completes everything successfully, the first landing mission (Artemis 3) may happen as soon as 2025.

ARTEMIS 2 LAUNCH DATE

The Artemis 2 launch date is provisionally scheduled for 2024, however that relies on the readiness of a few elements.

The Artemis 1 mission is slated for an uncrewed voyage around the moon in 2022, which will last more than a month. The mission will gather radiation data and engineering data to evaluate the SLS and

Orion spacecraft's fitness to transport people.

If Artemis 1 completes its mission successfully from launch to splashdown, Artemis 2 will be planned next. The crewed trip will need new spacesuits that are engineered to outlast the lunar orbiting environment, which has greater radiation than within low Earth orbit where humans have better protection.

As of August 2022, NASA is constructing essential elements of the Artemis 2 hardware with the expectation of performing testing and integration later in the process.

WHO WILL FLY ON ARTEMIS 2?

The Artemis 2 crew has not yet been named, but we do know that it will have four astronauts.

NASA announced in 2022 that all of its astronaut corps would be eligible for flights in the Artemis program. Additionally, the Canadian Space Agency will obtain an astronaut seat for this flight. The CSA has four astronauts available and has also not named its Artemis 2 crew member.

Canada earned its astronaut seat after agreeing to contribute robotics to NASA's human lunar mission. NASA aims to establish a space station, codenamed Gateway, that will reside in orbit around the moon to enable research and landing missions.

Canada's Canadarm3 will undertake repairs and maintenance on Gateway and will feature artificial intelligence to allow for some autonomous work.

The AI will be particularly critical while the station is uncrewed, as NASA expects pauses in between astronauts. This strategy is distinct from the International Space Station's constant human presence yet required for fiscal and logistical reasons.

WHAT WILL ARTEMIS 2 DO?

Artemis 2 will be the first significant test of the SLS and Orion spacecraft systems with people on board.

The mission will attempt to reach four primary indicators of readiness, according to the Canadian Space Agency: mission planning, system performance, crew interfaces, and guiding and navigation systems.

The mission plan will see Orion travel in an orbit known as "hybrid free return," which would see the spacecraft loop Earth twice to gather up speed for the trans-lunar injection. At the moon, Orion will also utilize lunar gravity as a "slingshot" to boost up speed for the return back to Earth, according to CSA.

The mission is planned to last between eight and 10 days, but may be extended to as long as three weeks depending on the mission goals, the agency noted.

The four astronauts on board Artemis 2 will be the farthest individuals to fly from Earth since 1970's Apollo 13, providing the current mission achieves its anticipated maximum altitude of 5,523 miles (8,889 km) beyond the moon's surface.

The European Space Agency claims the mission will need to achieve the following milestones:

Launch from NASA's Kennedy Space Center's Launch Pad 39B to low-Earth orbit.

A maneuver in Earth orbit to increase the perigee, or the lowest point of the orbit, around 40 minutes after liftoff. This will be conducted using the SLS Interim Cryogenic Propulsion Stage (ICPS).

A burn to increase the apogee, or highest section of the orbit, again utilizing the ICPS.

A system check at 42 hours after the mission starts to confirm the orbit is proper, ranging from (112 miles) to 185 km at the closest point to Earth and 1,616 miles (2,600 km) at its highest point.

The ICPS will be discarded and Orion will execute a translunar injection to go to the moon. The mission to the moon will take four days and have a maximum height of 5,523 miles (8,889 km) above the moon's surface.

The spaceship will return home. Once the spacecraft approaches Earth, the crew module will separate from the European Service Module and the crew module adapter, enabling a splashdown in the Pacific Ocean.

WHAT COMES AFTER ARTEMIS 2?

Investigators working with Artemis 2 will spend many months at the least evaluating data. The next mission in line is a landing mission named Artemis 3, which will arrive on the surface of the moon in 2025 if everything goes to plan.

NASA's Office of the Inspector General has raised concerns about such a schedule.

There have been delays in getting the human landing mechanism ready, which will employ SpaceX's Starship, owing to technical and legal concerns. In addition, there were development delays in the spacesuits that NASA was manufacturing; the government has shifted to private vendors to fill the gap.

Assuming Artemis 3 arrives on the surface in 2025, it would be the first landing mission by humans since NASA's Apollo 17 in 1972.

Frequently asked questions: Your curiosities acknowledged

What is the Artemis program?

Humans have not left Earth's orbit since Apollo 17 returned from the Moon in 1972.

NASA has been working to alter that since 2004 when then-U.S. President George W. Bush launched the Vision for Space Exploration, a program to bring people back to the Moon and ultimately to land on Mars.

Since then, NASA's deep space initiatives have had a variety of names: Constellation (2004-2010, aimed at lunar surface and Mars), Journey to Mars (2015-2018, targeted cislunar space, asteroid, and Mars), and Moon to Mars (2018 to present, targeting lunar surface and Mars).

Through its current Artemis program, NASA envisages sending men to the lunar south pole by 2025 and ultimately establishing a permanent presence on the Moon.

The initiative is a consequence of the Trump administration's Space Policy Directive 1 and a March 26, 2019 address by former Vice President Mike Pence instructing NASA to reach the Moon by 2024.

Artemis is meant to place people on the Moon swiftly and concentrate on Mars as a long-term human spaceflight objective after that.

The preliminary short-term proposal comprises employing both private rockets and NASA's Space Launch Vehicle, the Orion crew capsule, and a commercial lunar landing system. A modest space station in lunar orbit named the Gateway would service future surface expeditions.

The Planetary Society's principles for human spaceflight explain how we assess, support, and debate any proposals for human spaceflight.

Why is the program named Artemis?

Artemis is the legendary Greek goddess of the Moon and the twin sister of Apollo. The relationship with the expedition which first sent people to the Moon 50 years ago hence seems evident.

The crewed spaceship presently in construction meanwhile is dubbed Orion. Orion is one of the most known constellations in the sky, while in Classical mythology Orion is the hunting partner of Artemis.

What is the Space Launch System?

The Space Launch System (SLS) is a huge rocket based on Space Shuttle-derived technology. It is a bigger version of the Shuttle stack that swaps out the winged orbiter for either cargo or the Orion crew capsule on top.

The vehicle's core stage is a stretched Shuttle external fuel tank driven by four Space Shuttle (RS-25) main engines. (During the Shuttle program these engines were reconditioned and reused; for SLS they will be discarded in the ocean.) Assisting the core stage during the early phase of flight is a pair of five-segment Space Orion is a manned vehicle capable ofShuttle solid rocket boosters.

Orion
Accommodating up to four humans on deep-space voyages, similar in principle but

103

with a bigger cabin than the gumdrop-shaped Apollo capsules. Unlike capsules built primarily for transit to low-Earth orbit, Orion's heat shield can endure the high-velocity re-entry required for returning from deep space.

The Orion spacecraft includes three key components: a pressurized crew capsule, a service module, and a launch abort tower, which is nominally discarded during ascent.

Lunar Gateway

The Lunar Gateway is a tiny space station in lunar orbit that would operate as a fuel and supply store, a scientific outpost, and a waypoint for missions to and from the lunar surface.

The Gateway is presently not needed to be operational for the first 2025 Moon landing. NASA is soliciting private businesses to offer Gateway cargo transportation services, similar to the way it does for the International Space Station.

CAPSTONE

NASA will deploy a tiny spacecraft dubbed CAPSTONE (Cislunar Autonomous Positioning System Technology Operations and Navigation Experiment) in 2022 to the same lunar orbit Gateway will occupy.

The microwave oven-sized CubeSat will test out a variety of essential technologies necessary for Artemis, including spacecraft-to-spacecraft communication utilizing the Lunar Reconnaissance Orbiter.

Lunar Landers

NASA sought private businesses to create lunar lander systems that would ultimately dock with the Gateway. In April 2021, the space agency revealed it had picked

SpaceX's Starship to assist land people on the Moon.

A visiting Orion crew would board the lander, transport it to the surface, then return in either an ascent module or the complete vehicle. Early landers would only be capable of brief surface stays, whereas subsequent vehicles would be able to house workers through the lunar night.

Why is NASA going back to the Moon?

NASA is not only trying to recreate the exploits of the Apollo missions with Artemis, but rather to get to the Moon 'and remain there. That entails researching the prospect of creating outposts both in lunar orbit and on the Moon's surface, while the major aim for now still involves sending people to the Moon by the middle of the decade.

Key NASA mission goals include:

Equality: a primary target for NASA is to place the first woman and first person of color on the lunar surface.

Technology: from rockets to spacesuits, the technologies now being researched are aimed to pave the path for future deep-space expeditions.

Partnerships: the Artemis program is one of NASA's first large-scale collaborations with commercial companies, such as SpaceX and Boeing.

Long-term presence: where the Apollo 17 crew spent three days on the lunar surface, Artemis aims to establish a base to extend the trips to weeks and possibly months.

Knowledge: while more is known about the Moon compared with 50 years ago (and technology has substantially evolved), NASA says that this next series of missions will be able to gather samples more strategically than during the Apollo period.

Resources: the finding of water on the Moon and probable reserves of rare minerals offer promise for both scientific and commercial research and utilization.

Has the Artemis launch been delayed?

NASA's first stated objective was to put people on the Moon by 2024. The agency said in November 2021 however that this date will be moved out to no sooner than 2025.

Even this timetable is far from guaranteed, with NASA Inspector General Paul Martin indicating that the crewed lunar landing would likely slide until 2026 at the earliest.

The landing is the third of a multi-year mission commencing with Artemis I, an uncrewed voyage around and beyond the Moon. Each of these missions has ambitious deadlines of its own.

Will traveling to the Moon help humanity settle on Mars?

While the voyage to the Moon takes three days, reaching Mars is a significantly longer and more challenging endeavor. NASA envisions Artemis as establishing the basis for both international space agencies and private firms to construct a lunar population and economy, and from there someday take people to Mars.

What about others trying to go to the Moon?

NASA is not the only one wanting to get to the Moon. Russia, China, and Japan have all proposed sending their astronauts to the Moon, with programs at varying degrees of completion.

But space flight has now transitioned into an age of private exploration. One of the top businesses is SpaceX, which wants to launch a Moon mission as soon as next year.

This is envisaged employing its Starship spaceship atop a booster rocket known as Super Heavy. Unlike NASA's SLS, SpaceX's launcher and spacecraft are meant to be reusable, which will dramatically decrease the price of each trip.

While the Starship has successfully flown and landed from high altitude, it has yet to be tested in flight with the Super Heavy rocket, or orbit around the Moon. Once these tests are complete, it will clear the way for its planned 2023 lunar mission, entitled dear moon, to launch.